STEMD² Research & Development Group

Center on Disability Studies

College of Education

University of Hawaiʻi at Mānoa

http://stemd2.com/

Printed in the United States of America. First published (2017) by the Center on Disability Studies, University of Hawaiʻi at Mānoa, Honolulu, Hawaiʻi. This book is based upon work supported by the Department of Education, Native Hawaiian Education Act Program under award #S362A140018. Any opinions, findings, and conclusions or recommendations expressed in this material are those of the authors and do not necessarily reflect the views of the United States Department of Education. For further information about this book and the Nāʻepapa Ka Hana project, please contact Dr. Kaveh Abhari at abhari@hawaii.edu.

ISBN: 978-0-9983142-3-5

Second release, 2020

Ne'epapa Ka Hana Eighth-Grade Mathematics Resources

Let's Play the 'Ukulele

Student Activities

Project Directors	Kaveh Abhari Kelly Roberts
Content Developers	Robert G. Young Justin Toyofuku
Publication Designers	Katie Gao Robert G. Young Lauren Ho MyLan Tran
Content Contributors	Lock Lynch Luanna Peterson Holm Smidt Remy Pages Kimble McCann

Acknowledgments

We would like to thank Ruth Silberstein, Cody Kikuta, Katy Parsons, Crystal Yoo, Kelli Ching, Jacob Koseki, and Chase Cabana for advising on middle school mathematics education. Also, thank you to Keola Nakanishi and Mana Maoli, who were key motivators and supporters in the development of *Let's Play the 'Ukulele*.

Suggested Citation

Let's Play the 'Ukulele (2020). Ne'epapa Ka Hana Eighth-grade Mathematics Resources. STEMD^2 Book Series. STEMD^2 Research and Development Group. Center on Disability Studies, University of Hawai'i at Mānoa. Honolulu, Hawai'i.

Contents

Preface

About the STEMD² Book Series

The STEMD² Book Series for eighth-grade mathematics was developed as part of a technology-enabled pedagogical approach (Ne'epapa Ka Hana model) for teaching mathematics in Hawai'i middle schools. This book series provides Hawai'i middle school teachers resources and training to incorporate problem-based learning, social learning and inclusive pedagogy through a culturally relevant mathematics curriculum. The series currently includes:

> **Let's Build a Canoe** – Student Activities and Teacher's Guide (Common Core aligned)
> **Let's Play the 'Ukulele** – Student Activities and Teacher's Guide (Common Core aligned)
> **Let's Go Fishing** – Student Activities and Teacher's Guide (SBAC aligned)
> **Let's Make Da Kine / E Hana Kākou** – Student Mini Projects and Teacher's Guide in English and 'Ōlelo Hawai'i (Skill development)

The printed and online resources produced by NKH through STEMD² are fully aligned with the Common Core Standards for Mathematical Practice and Content and the Smarter Balanced Assessments for mathematics. Based on the GO Math!® curriculum structure, the NKH STEMD² book series and social learning platform (`community.stemd2.com`) is flexible for teachers to implement partially or fully in their classrooms, as a tool to encourage students' interest and achievement in STEM subjects.

Funded by a three-year grant from the Department of Education's Native Hawaiian Education Act Program, Ne'epapa Ka Hana is a project of the STEMD² Research & Development Group in the Center on Disability Studies at the University of Hawai'i at Mānoa. More information about STEMD² is available online at `www.stemd2.com`.

Unit 1: Real Numbers, Exponents, and Scientific Notation

Activity 1.1 - The Meaning of Sound

When a rock falls into water, the water gets pushed and pulled around, creating little waves that spread out. This also happens in air, which is how we hear *sound*. When something like a 'ukulele string moves (vibrates), it squishes and stretches the air around it. This squishing and stretching of air then spreads around as a *sound wave*. When the air squishes and stretches quickly and the wave reach our ears just right, we call this *sound*.

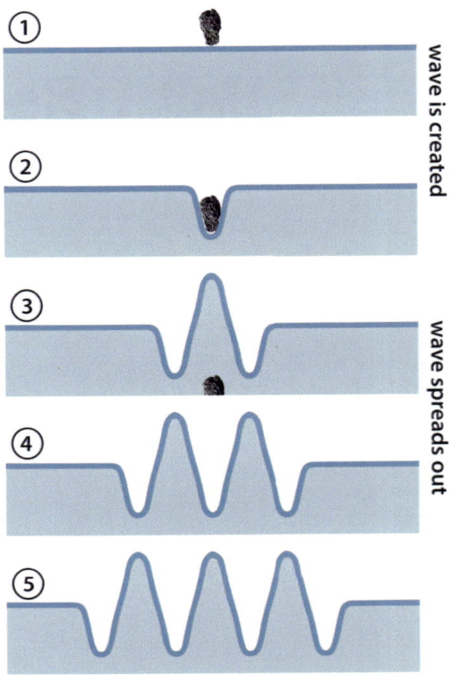

The *frequency* of a sound describes how often the air squishes and stretches in one second, and its unit is called a hertz (Hz). Hypothetically, if you could hear a sound of 1 Hz, then your ears are noticing that the air is squished and stretched 1 time each second. If you could hear 9 kHz (kilohertz), then your ears are noticing that the air is squished and stretched 9,000 times per second. In music, some specific frequencies are called *notes*, and we give them names like B, C flat, and F sharp.

You might have heard people talk about sounds that are "high pitch" or "low pitch." Like frequency, *pitch* describes how many times a sound wave is squished and stretched in one second (measured in Hz), which is why we hear different "high" and "low" sounds. We can say that *frequency* describes what the wave is physically doing, and *pitch* is how this frequency feels to our ears. In this book, we use the scientific term *frequency* to talk about the sound wave.

1. Ask your teacher to play this video, which helps us visualize the sound waves while playing different frequencies (`http://bit.ly/1L6ZYye`). Cover your ears a little if it gets irritating. (You can search "Hearing Test" on YouTube for the same video).

a. What is the lowest and highest frequency that you can hear?

b. Why do you think people have different answers to Part a?

c. Some of the frequencies can only be heard with the correct sound equipment. If you can't hear some of the frequencies, is it because of the sound equipment? You need to discuss with your classmates to figure this out.

2. Explain what the squiggly line in the video is showing you. What are some interesting things that you notice about it? Talk about this with your group. Share with the classroom or on the online comment section.

Activity 1.2 - Octaves I

1. Let's look at how different kinds of numbers affect sound.

a. What are whole numbers and what are integers? It will be easier to remember if you describe it in your own words instead of copying from a textbook.

b. Name three integers that are not whole numbers.

2. When the frequencies are multiples of each other by a power of two, we call them *octaves*. For example, 3 Hz and 24 Hz are octaves, because you can multiply 3 Hz by a power of two ($8 = 2^3$) to get 24 Hz. 3 Hz and 25 Hz are not octaves.

Which of the following frequencies are octaves? Write as pairs and write the lower frequency first. For example: 3 Hz & 24 Hz.

30 Hz
32.70 Hz
60 Hz
49.05 Hz
40 Hz
50 Hz
16.35 Hz

3. Here are two graphs of three notes (frequencies) played together. In the first graph, all three notes are octaves of each other. In the second graph, the three notes are not octaves.

Three sounds that are octaves

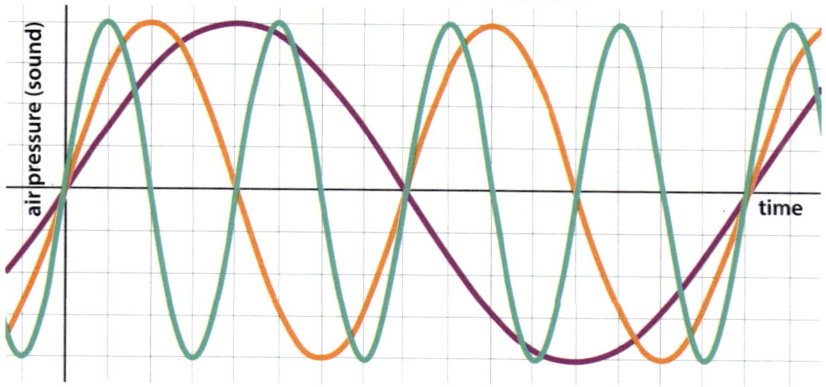

Three sounds that are *not* octaves

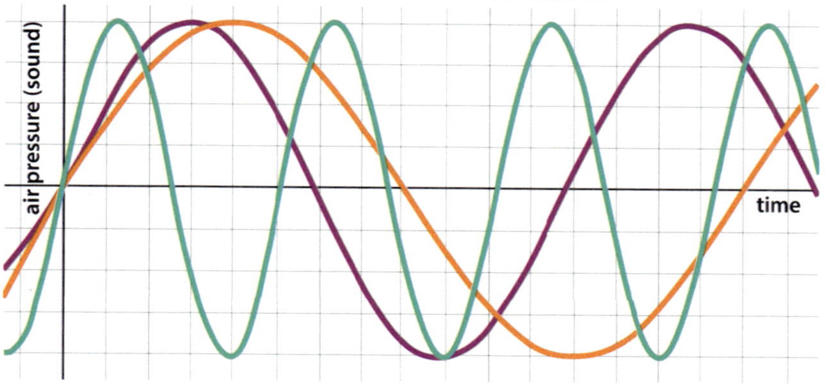

When octaves are played together, they sound really nice and sound like one note. In fact, we even give these notes the same name! We'll talk about this in Activity 1.3.

a. Compare the two graphs. What are some things that they have in common? What are some differences?

b. The notes in the first graph sounds nicer than the second graph. Why do you think that is?

c. The notes in the first graph have the same name, but the notes in the second have different names. Why do you think this makes sense or is practical to do?

d. Share your answers from Parts a-c with your class or in the online comment section !

Activity 1.3 - Octaves II

Notes are specific frequencies that are named with the letters A to G (for example, B, C flat, and F sharp), which we use when playing or reading music. Recall that octaves are notes with frequencies that are powers of two multiples of each other: $2^1 f, 2^2 f, 2^3 f$, etc. So if a note has frequency f, then its octaves are $2f, 4f, 8f$, etc. We say that $2f, 4f$, and $8f$ are one, two, and three octaves higher than f.

We can also have negative powers of two: $2^{-1}f, 2^{-2}f$, and $2^{-3}f$. Therefore, $f/2$, $f/4$, and $f/8$ are also octaves, which are one, two, and three octaves lower than f.

Notes that are octaves apart have the same name. So for example, if F♯ (F sharp) has frequency f, then the next time that F sharp shows up is on its next octave, with frequency $2f$.

1. What frequency is 2 octaves higher than 250 Hz?

2. What frequency is 5 octaves higher than 300 Hz?

3. What frequency is 4 octaves lower than 4800 Hz?

4. In a 12-note scale (called a *chromatic scale*), an octave is divided into 12 frequency steps. Some notes have two names like C♯ (C sharp) and D♭ (D flat). Why do you think this is?

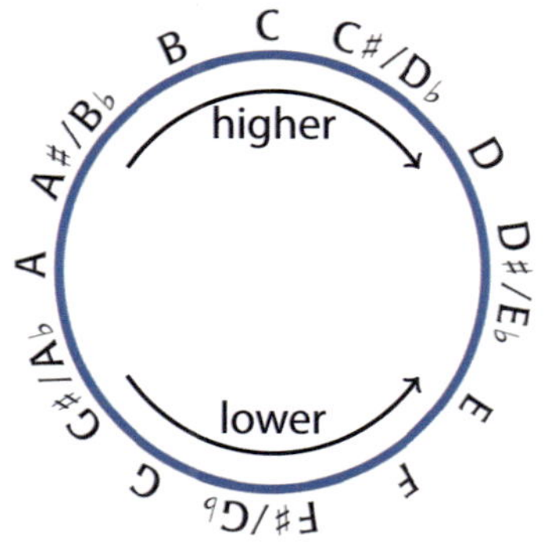

In the next activity, we will find all the notes between f and $2f$ using rational numbers. We will use a method called Pythagorean tuning. Later, we'll talk about another type of tuning.

Activity 1.4 - Notes of a Scale I

Did you notice from Activity 1.2 that the shape of a wave repeats? The *period* of a wave describes how long it takes for a wave to repeat. In a sound wave with frequency of 0.5 Hz, air squishes and stretches 0.5 times every second so we have a period of 2 seconds.

Sound wave frequency of 0.5 Hz

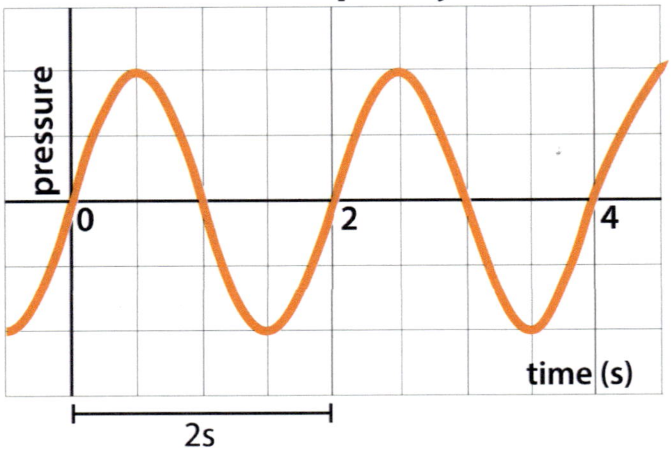

In a sound wave with frequency 1 Hz, air squishes and stretches 1 time every second so we have a period of 1 second.

Sound wave frequency of 1.0 Hz

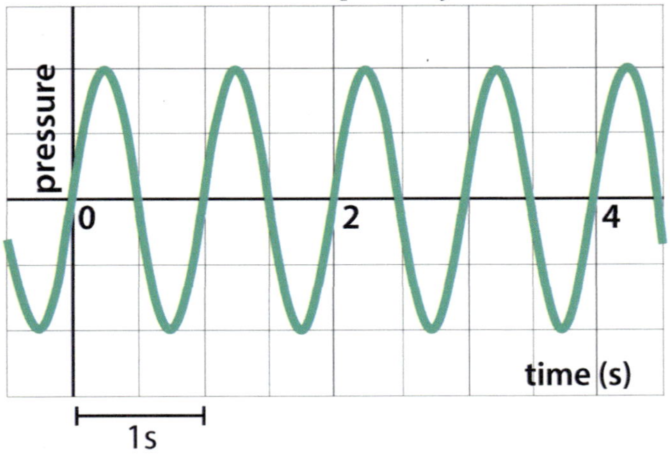

0.5 Hz and 1 Hz are octaves and sound so nice when played together that we give them the same name. Look at this graph of the two notes played together. They sound nice because *two* of one note fits nicely in *one* of the other note. In fact, notes sound nice if they are a ratio of small integers to each other, and sometimes sound awful if they are not small ratios to each other.

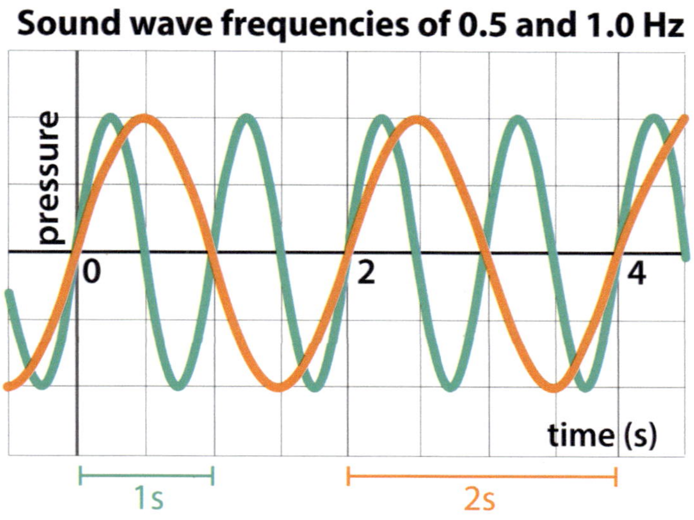

The ratios of small integers 1/2 and 2/1 helped us find octaves. The ratios 2/3 and 3/2 will help us find all the other notes within one octave.

1. This activity will be easier if we know our powers of 2 and 3. Fill out Table 1.1.

Table 1.1: Activity 1.4 part 1. Powers of 2 and 3.

n	2^n	3^n
1	2	3
2	4	9
3		
4		
5		243
6		
7		
8		
9	512	
10		

2. f is the first note. To find five more we multiply a note by 3/2 to increase the frequency. Remember, we want the notes between $1f$ and $2f$ because that is the range of one octave. So if our note gets too high when we multiply by 3/2, then we have to lower the note one octave by multiplying by 1/2. Complete this guide (you can use Table 1.1, if you like):

a. **First note:** f **Multiply by 3/2:** $(3/2)f$
If note is too high, lower it one octave by multiplying by 1/2. **New note:** $(3/2)f$

b. **Previous note:** $(3/2)f$ **Multiply by 3/2:** $(9/4)f$
If note is too high, lower it one octave by multiplying by 1/2. **New note:** $(9/8)f$

c. **Previous note:** $(9/8)f$ **Multiply by 3/2:**
If note is too high, lower it one octave by multiplying by 1/2. **New note:**

d. **Previous note:** **Multiply by 3/2:**
If note is too high, lower it one octave by multiplying by 1/2. **New note:**

e. **Previous note:** **Multiply by 3/2:**
If note is too high, lower it one octave by multiplying by 1/2. **New note:**

3. To find the last six notes, we start with our first note f. Then we multiply a note by 2/3 to decrease the frequency. Remember, we want the notes between $1f$ and $2f$ because that is the range of one octave. So if our note gets too low when we multiply by 2/3, then we have to raise the note one octave by multiplying by 2. Complete this guide (you can use Table 1.1 if you like):

a. **First note:** f **Multiply by 2/3:** $(2/3)f$
If note is too low, raise it one octave by multiplying by 2. **New note:** $(4/3)f$

b. **Previous note:** $(4/3)f$ **Multiply by 2/3:** $(8/9)f$
If note is too low, raise it one octave by multiplying by 2. **New note:** $(16/9)f$

c. **Previous note:** $(16/9)f$ **Multiply by 2/3:**
If note is too low, raise it one octave by multiplying by 2. **New note:**

d. **Previous note:** **Multiply by 2/3:**
If note is too low, raise it one octave by multiplying by 2. **New note:**

e. **Previous note:** **Multiply by 2/3:**
If note is too low, raise it one octave by multiplying by 2. **New note:**

f. **Previous note:** **Multiply by 2/3:**
If note is too low, raise it one octave by multiplying by 2. **New note:**

Activity 1.5 - Notes of a Scale II

1. With your group, check your answers to Activity 1.4 by confirming that you found all the notes in Table 1.2 on the next page. In the examples, we gave you the first note f, the next octave $2f$, and the following four notes:

$\frac{3}{2}f$, $\frac{9}{8}f$, $\frac{4}{3}f$, and $\frac{16}{9}f$

If you are missing some notes, work with your group to figure out where you might have made a mistake. Use teamwork!

2. What are some patterns and interesting features (or annoying ones) that you found from filling out the table? Discuss with your group and share with your classroom or on the online comment section.

3. These twelve notes are called the *chromatic scale*!

It turns out that the twelve notes of the chromatic scale have a nice pattern. If f is the frequency of a note, then the next note on the chromatic scale is either

$\frac{2^8}{3^5}f$ or $\frac{3^7}{2^{11}}f$

Look back at the Table 1.2 to check if this is true!

Table 1.2: Frequencies of a scale (Activity 1.5.1)

Order of note (lowest to highest)	Frequency (as a fraction)	Frequency (as a fraction of exponents)	Found it?
1st	f	f	✓
2nd	$\frac{256}{243}f$	$\frac{2^8}{3^5}f$	
3rd	$\frac{9}{8}f$	$\frac{3^2}{2^3}f$	✓
4th	$\frac{32}{27}f$	$\frac{2^5}{3^3}f$	
5th	$\frac{81}{64}f$	$\frac{3^4}{2^6}f$	
6th	$\frac{4}{3}f$	$\frac{2^2}{3}f$	✓
7th	$\frac{1024}{729}f$	$\frac{2^{10}}{3^6}f$	
8th	$\frac{3}{2}f$	$\frac{3}{2}f$	✓
9th	$\frac{128}{81}f$	$\frac{2^7}{3^4}f$	
10th	$\frac{27}{16}f$	$\frac{3^3}{2^4}f$	
11th	$\frac{16}{9}f$	$\frac{2^4}{3^2}f$	✓
12th	$\frac{243}{128}f$	$\frac{3^5}{2^7}f$	
13th (next octave)	$2f$	$2f$	✓

Activity 1.6 - Notes of a Scale III

The way that you made the *chromatic scale* in Activity 1.5 is called **Pythagorean Tuning**. Instruments like the piano and keyboard use chromatic scale that is made in a slightly different way. Instead of Pythagorean tuning, these instruments use **equal temperament tuning**. In equal temperament tuning, the frequency of each note is always a number α times the last note f. So if f_1, f_2, f_3 are the first, second, and third notes, then $f_2 = \alpha f_1$, $f_3 = \alpha f_2 = \alpha^2 f_1$.

1. What is the difference between Pythagorean tuning and even temperament tuning?

2. Write the seventh note, f_7 in terms of α and f_1.

3. The thirteenth note, f_{13} is $\alpha^{12} f_1$. Check that everyone in your group understand this.

$$\begin{aligned} f_2 &= \alpha f_1 \\ f_3 &= \alpha f_2 &&= \alpha^2 f_1 \\ &\ldots \\ f_{13} &= \alpha f_{12} &&= \alpha^{12} f_1 \end{aligned}$$

The thirteenth note is also the next octave $2f_1$. So α is the twelfth root of two.

$$\begin{aligned} f_{13} &= 2f_1 \\ \alpha^{12} f_1 &= 2f_1 \\ \div f_1 & \quad \div f_1 \\ \alpha^{12} &= 2 \\ \sqrt[12]{\alpha^{12}} &= \sqrt[12]{2} \\ \alpha &= \sqrt[12]{2} \end{aligned}$$

Compare the number $\sqrt[12]{2}$ from equal temperament with $\frac{2^8}{3^5}$ and $\frac{3^7}{2^{11}}$ from Pythagorean tuning. Write them from least to greatest. Are they close together or far apart? You may use a calculator for this.

4. Assume that the first note, f_1 is 1 Hz. Without a calculator, write the seventh note as a decimal number. Show your work and round to 2 decimal places.
Hint: In part 2 you wrote an equation for f_7 and in part 3, you have α.
Hint: $(\sqrt[b]{x})^a = x^{\frac{a}{b}}$. So for example, $\left(\sqrt[3]{12}\right)^6 = 12^{\frac{6}{3}} = 12^2 = 144$.

5. What are *rational numbers* and *irrational numbers*? Describe in your own words.

6. Does the Pythagorean tuning use only rational numbers, only irrational numbers, or all real numbers? What about even temperament?

Activity 1.7 - Frequency Ranges

Let's go back to the video that plays and shows different frequencies (`http://bit.ly/1L6ZYye`). Ask your teacher to play it in the classroom. (Cover your ears a little if it gets irritating.)

Answer the following questions in scientific notation.

1. What are the lowest and highest frequencies that you can hear from this video?

2. Use the internet to find out what's the highest frequency that a 'īlio (dog) can hear?

3. The highest note that a 'īlio can hear is how many times higher than the highest note that you can hear?

4. The Hawaiian spinner dolphins are called nai'a. The highest note that a nai'a can hear is about 150 thousand hertz. This is how many times higher than the highest note that you can hear?

5. The speed of sound in air is about 343.2 meters per second. This is about 22.9% of the speed of sound in seawater. How does the speed of sound affect how and what you hear? How do you think this affects the lives of seals, turtles, and other animals that are always in and out of the water? Discuss with your classroom on the online comment section.

Unit 2: Proportional and Nonproportional Relationships and Functions

Activity 2.1 - The Speed of Sound

Last time, we talked about the speed of sound. Let's round the speed of sound in air to 340 meters per second.

1. If we make a sound at position $y = 0$ meters at time $x = 0$ seconds, write an equation for y, where the sound wave is at time x.

2. Graph your equation from part 1. For the x axis, plot x from 0 seconds to 4 seconds. Make sure that you choose an appropriate y axis that graphs your line nicely. If the y range is too short, then the graph will fly off the top of the page right away. If the y range is too long, then the line will be really flat and may be hard to read. If it is okay with your teacher, you can use `www.desmos.com/calculator`.

3. Suppose that you woke up late for hula practice and you're 1000 meters away from your kumu. You see that your kumu hits her ipu to the ground at time x = 0 seconds. How many seconds later do you hear the sound? Use your graph from part 2 to help you.

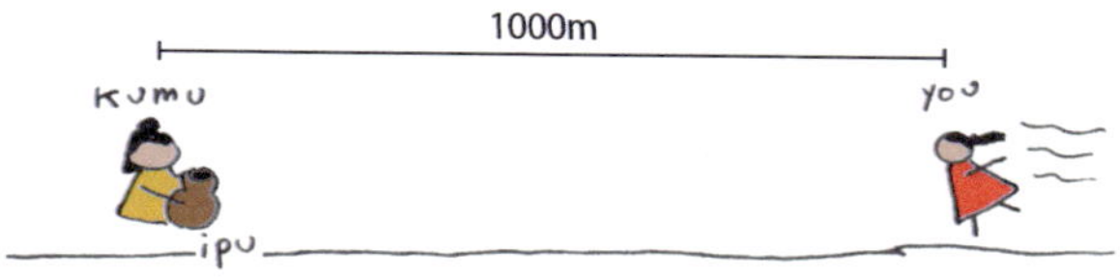

4. Sometimes we hear a sound, then it bounces off of something and we hear it again. For example, if your kumu is sitting in front of big rock, 300 meters away, she will hear her ipu when she hits it. Then she hear it a second time when the sound wave travels 300 meters to the rock then bounces back another 300 meters to her. This is called an *echo.*

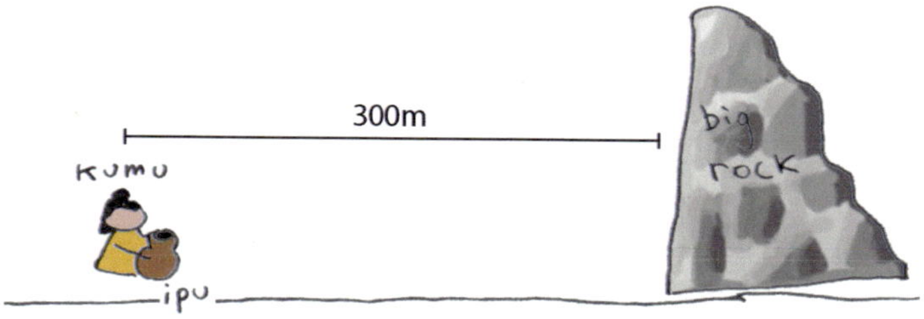

How much time passes between the first and second time that your kumu hears her ipu? Use your graph from part 2 to help you.

5. Suppose that you sat down between your kumu and the big rock.

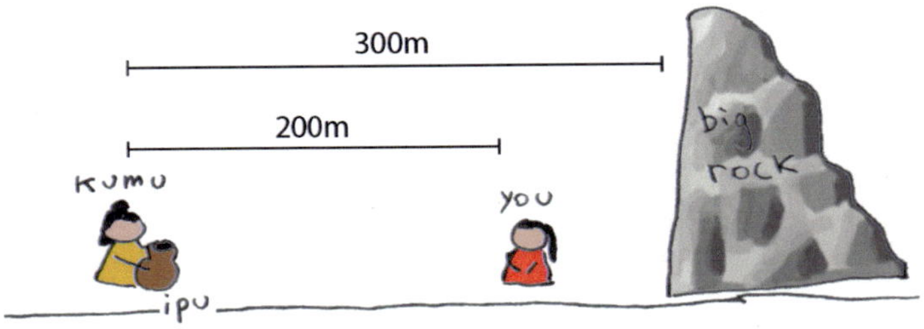

How much time passes between the first and second time that you hear the ipu? Use your graph from Part 2 to help you.

6. Sometimes echoes can really ruin a performance. How do you think we can get rid of echoes? For example, can you just change where you are sitting? Share your ideas with your classroom on the online comment section.

Activity 2.2 - Song Structure

Let's learn about song structure. Most songs are broken into smaller sections such as intros, verses, and choruses (hui). The song structure describes how the sections come together to make the song. For example, many songs have the structure: intro, verse, hui, verse, hui, and hui again.

Let's take a look at the parts of our state song, Hawai'i Pono'ī, written by King David Kalākaua.

Song structure	Hawaiian	English
Verse 1	Hawai'i pono'ī Nānā i kou mō'ī Kalani ali'i Ke ali'i	Hawai'i's own true son Be loyal to your king Your country's liege and lord The chief
Verse 2	Hawai'i pono'ī Nānā i nā ali'i Nā pua kou muli Nā pōki'i	Hawai'i's own true sons Look to your chiefs The flowers (children), your youngest ones The young
Verse 3	Hawai'i pono'ī E ka lāhui ē 'O kāu hana nui E ui ē	Hawai'i's own true sons People of loyal heart The only duty lies List and abide
Hui	Makua lani ē Kamehameha ē Na kaua e pale Me ka ihe	Father above us all Kamehameha e Who guarded in the war With his spear

1. Hawai'i Pono'ī usually takes 1 minute and 20 seconds to sing. This is because the song structure usually only contains the first verse, followed by the hui twice. During important Hawaiian events like the Merrie Monarch Festival, the song lasts about 2 minutes and 20 seconds. What do you think is the song structure of Hawai'i Pono'ī for important events? You can look online and/or listen to YouTube videos to see other song structures for this song.

2. Suppose that you and your classmates are singing a song called Mele Laina for different community events. Your class has to change the song structure to fit the different events, so your class plans to play the first few minutes of the Mele Laina, then repeat some sections x number of times until the end of the song. The hui is 40 seconds long and the verses are 1/3 minutes long.

a. Let y be the length of the performance (in minutes). If your class sings 5 minutes of Mele Laina, then repeats the hui x times, what is the equation for y?

b. Let y be the length of the performance (in minutes). If your class sings 6 minutes of Mele Laina, then repeats the verse x times, what is the equation for y?

c. Let y be the length of the performance (in minutes). If your class sings 4 minutes of Mele Laina, then repeats the entire 4 minute section x times, what is the equation for y?

d. Let y be the length of the performance (in minutes). If your class sings 6 minutes of Mele Laina, then repeats the both the verse and the hui section x times, what is the equation for y?

3. Graph your equations from Part 2. Ask your teacher if you may use DESMOS for this part.

4. Another class will be dancing for 10 minutes, so your class must sing Mele Laina for 10 minutes. Which of the four song structures from Part 2 would be best for this show and why?

Activity 2.3 - String Instruments I

When you pick a 'ukulele string, it vibrates, squishing and stretching the air around it to make a sound. The frequency that the string vibrates at depends on what the string is made of, how tight the string is, and how long the vibrating part of the string is. For most songs, we only change the length of the vibrating part of the string (by holding down the string at different points).

The length of the vibrating part of the string is inversely proportional to the frequency of the sound created. This means that if the length is decreased by a rate, then the frequency is increased by the same rate.

For example, suppose a string of length L makes a sound of frequency f.

Example 1:

If we shorten the length to $(1/2)L$ by pushing the string down in the middle, then the frequency increases to $2f$.

Example 2:

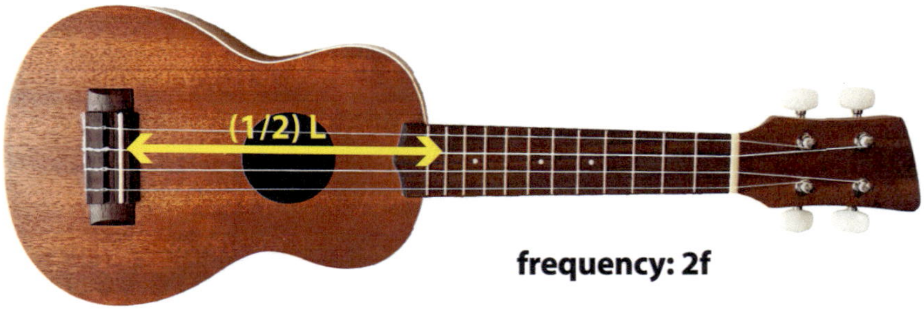

If instead, we change the length to $(2/3)L$, then the frequency changes to $(3/2)f$.

Example 3:

1. If x is the length of the vibrating string, and y is the frequency of the resulting sound, is the relationship between x and y linear or nonlinear? If it is linear, is it proportional or nonproportional? Explain. Make sure that everyone in your group is clear and agrees.

2. The note played on the string in Example 1 is called C. What is the name of the note in Example 2? (Hint: we talked about this in Unit 1).

3. The frequency of the note played in Example 1 is 261.6 Hz. What is the frequency of the note in Example 3?

4. Suppose that x is the length of a vibrating string, and y is the frequency of the resulting sound. Look at the following graphs. Discuss with your group if the graph correctly shows x and y. If the graph does not show the correct relationship between x and y, explain why not.

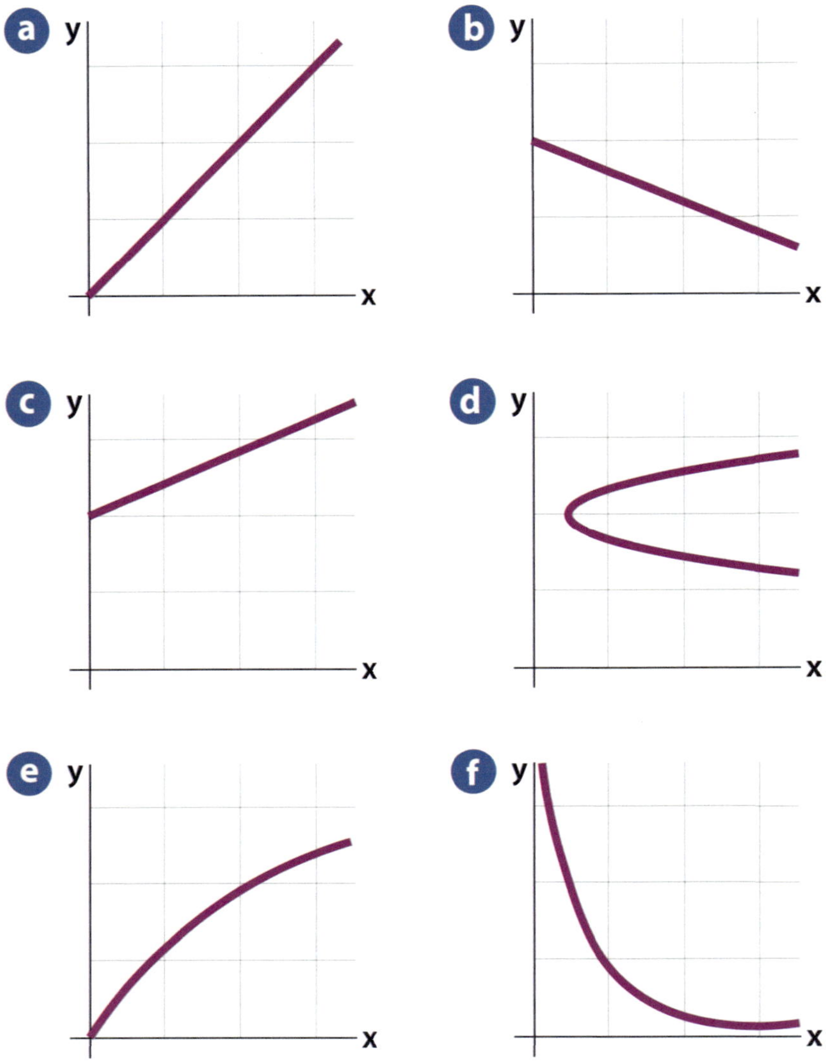

Activity 2.4 - String Instruments II

The strings on a 'ukulele are usually chosen and tightened so that they play the notes G (392 Hz), C (261.6 Hz), E (329.6 Hz), and A (440 Hz). As we learned earlier, we can also play other notes by shorting the length of the vibrating string. We do this by using our fingers to hold down the string at certain frets.

1. We can even find multiple ways to play the same note. For example, we can play the A note (440 Hz) by (1) picking the G string while holding the second fret, (2) picking the C string while holding fret 9, (3) picking the E string while holding fret 5, (4) and picking the A string without holding any frets.

a. Suppose that you were playing the four notes and ended up with this mapping diagram. Is "sound frequency" a function of string and fret? Why or why not? What is this mapping diagram is showing? What does it say about the 'ukulele in real life?

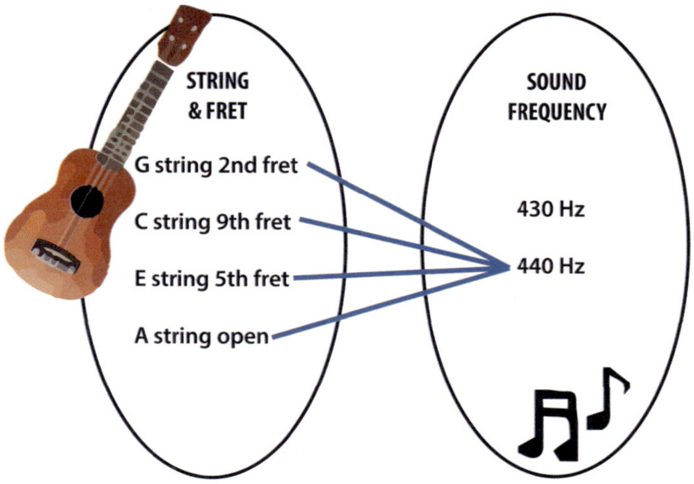

b. Suppose that you were playing the four notes but ended up with this mapping diagram instead. Does it show a function? What does it say about your ʻukulele this time?

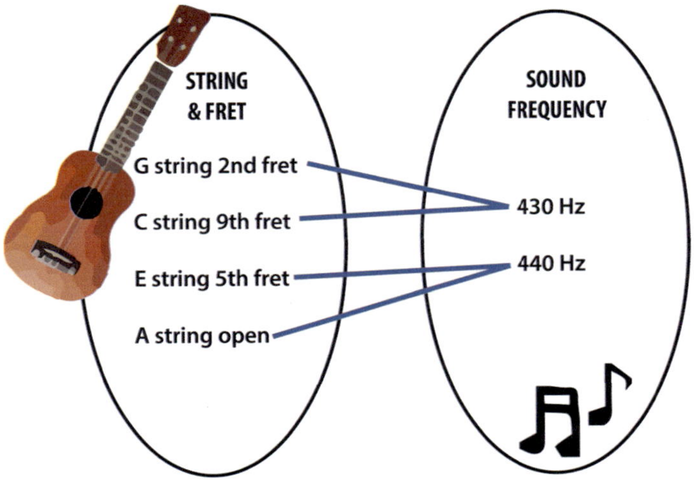

2. Cousins Kekoa, Kalani, Koa, Ku'uipo, Kau'i, and Kaimana are setting up their ʻukulele for a concert. They each play a string and a fret (left) and listen for the frequency (right).

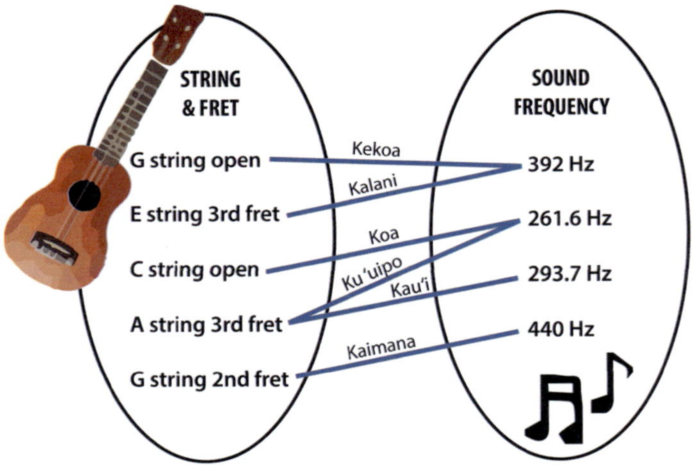

a. Which pair or pairs of cousins are playing the same note?

b. Which cousin probably needs to fix the way their 'ukulele is set up? There may be multiple answers. Feel free to discuss with your group and share with the class!

c. This mapping diagram actually does not show a function from "string and fret" to "sound frequency". Why not?

d. Who should stop playing in order to make this a function?

Unit 3: Solving Equations and Systems of Equations

Activity 3.1 - Stage Setup I

There are many different kinds of microphones. Choosing the right one depends mostly on what and where are you recording. If you want to record a sound coming from one direction and block out sounds from all other directions, then you might want to use a unidirectional microphone. These microphones clearly record what's right in front of it and ignores the stuff that's not in front of it.

Let's help set up the stage for a Hawaiian Choir of nine singers. On the y-axis are our three unidirectional microphones at the following points.
Mic A: (0, 4)
Mic B: (0, 8)
Mic C: (0, 13)
The Hawaiian Choir is split into three groups of three at the following points.
Group 1: (7, 4)
Group 2: (5, 8)
Group 3: (6, 12)

1. Plot these six points on a graph.

2.a. Microphone A is pointed directly at Group 1. Draw a line connecting these two points. What is the equation of the line that goes through these two points? (Use the point slope form: $y = mx + b$).

b. Microphone B is pointed directly at Group 3. Draw a line connecting these two points. What is the equation of the line that goes through these two points? (Use the point slope form: $y = mx + b$).

c. Microphone C is pointed directly at Group 2. Draw a line connecting these two points. What is the equation of the line that goes through these two points? (Use the point slope form: $y = mx + b$).

3. Solo! One of the singers has to perform a solo.

a. Where could they stand in order for microphones A and C to record them clearly? Use your answers from Part 2, and the substitution method.

b. Where could the soloist stand in order for microphones B and C to record them clearly? Use your answers from Part 2, and the elimination method.

c. Does it make more sense to stand at the point from Part 3a or from Part 3b? Why?

Activity 3.2 - Stage Setup II

Suppose that you set up four unidirectional microphones along the y-axis. The microphones are pointed left along the lines given by the following equations.
Mic A: $y = -2 + (2/3)x$
Mic B: $y = 2$
Mic C: $y = 6 + (2/3)x$
Mic D: $y = 8$

1. Graph the four equations for the mic directions. Label where the microphones are on the y-axis.

2. The soloist should stand where two microphones intersect. What is the coordinate where the soloist should stand? There is more than one possible answer, why did you choose to put the soloist here?

3. The backup singers should also stand where two microphones intersect, but away from the soloist. What are the coordinates where they should stand? There is more than one possible answer, why did you choose to put the backup singers here?

4. There is a 'ukulele player standing at (3, 5). You have to set up one more microphone on the y-axis for their 'ukulele. This microphone is specially set up to record the sounds of a 'ukulele but microphones B and D are set up to record voices. What is the equation of the line that goes from the microphone to 'ukulele, but NEVER intersects microphones B or D?

Unit 4: Transformational Geometry

Activity 4.1 - Semitones I

A *semitone* is the difference between two notes on a scale. Many instruments can play 12 semitones within an octave. On a 'ukulele, each fret is one semitone apart. When you move your finger one fret closer to the sound hole, the frequency goes up one semitone. When you move your finger one fret further from the sound hole, the frequency goes down one semitone.

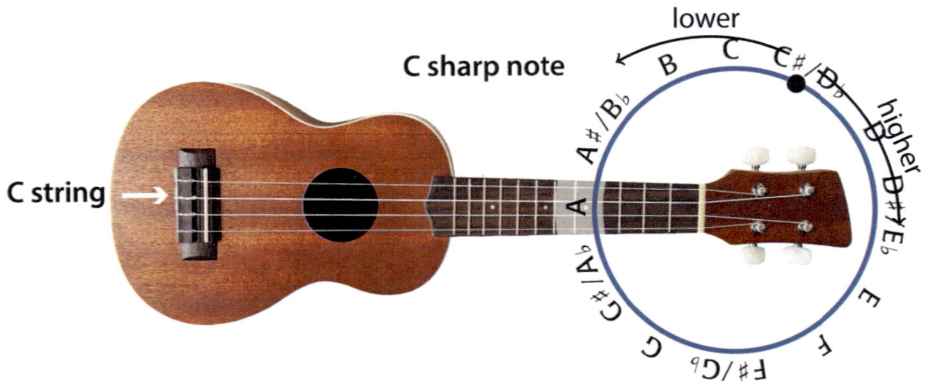

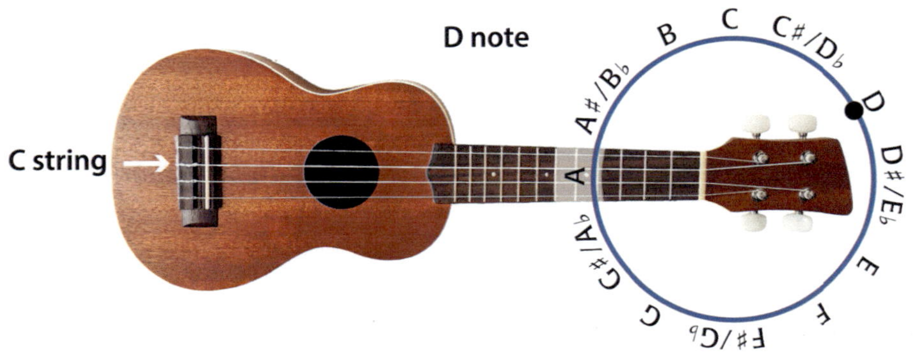

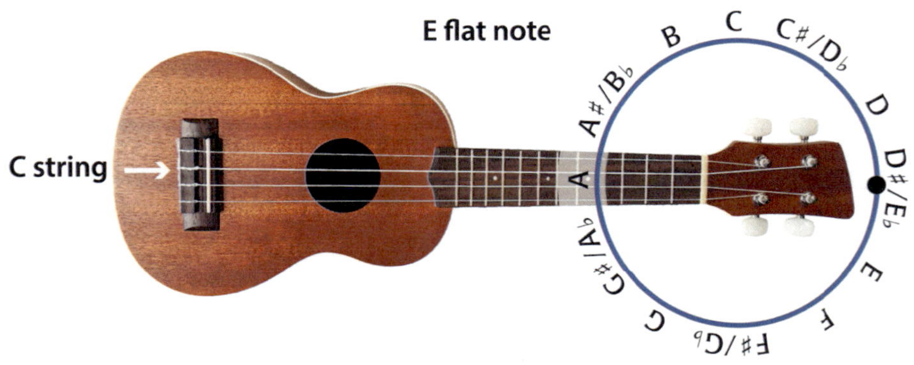

1. From the E♭ (E flat) in the last example, how many frets towards the sound hole do we have to move to play the next G note?

2. What are two ways to play the C note on this string? What's the difference between these two ways?

3. If you play the four strings (G, C, E, and A) open—without holding down any frets—then the C string will vibrate at the lowest frequency. Can a 'ukulele ever play a note lower than this note? If so, how? If not, why not?

4. How many octaves do you think can be played on a 'ukulele? Why? Share your thoughts with your class or on the online comment section.

Activity 4.2 - Semitones II

Use this chromatic scale circle to help you with the following parts.

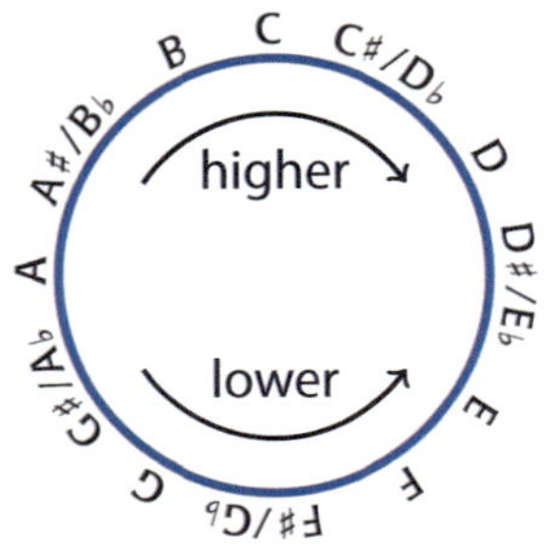

Here's our 'ukulele fretboard with a grid. Notice that the E string is on line $y = 5$. The third fret is on the line $x = 9$.

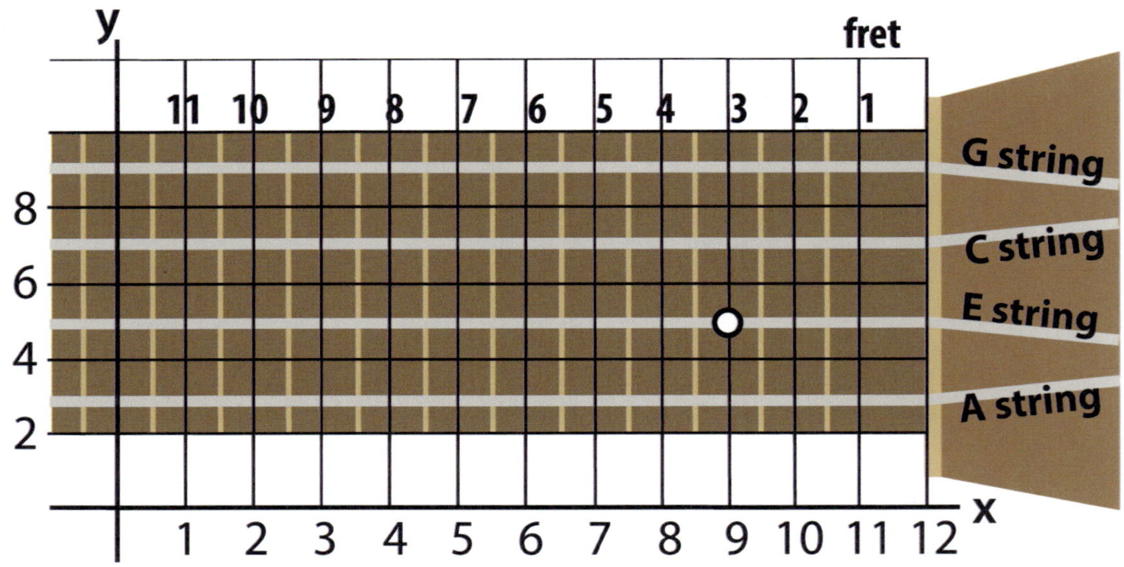

1. If we place a finger on (9, 5) and play the E string, then we'll hear the G note. How can you use the chromatic scale circle to show this?

2. Suppose that we want to play the lowest A note on the E string. It is at $(9+\alpha, 5+\beta)$. What is the value of α and β?

3. The lowest A note on the C string is on at $(9+\alpha, 5+\beta)$. What is the value of α and β?

4. Suppose that you're playing a note on (x, y). You want to play another note at $(x+\alpha, y+\beta)$. If the second note is on the same string and 3 semitones higher than the first, what is the value of α and β?

5. Suppose that you're playing a note on (x, y). You want to play another note at $(x+\alpha, y+\beta)$. If the second note is on the same string and 2 semitones lower than the first, what is α and β?

Activity 4.3 - Chords

Let's learn about chords. A chord is a group of three or more notes that are heard together. Sometimes, the notes of a *chord* follow a special pattern, which has a special name.

A *major chord* is made up of three notes. The lowest note is called the *root*. The second note is called the *major third* and is four semitones higher than the root. The last note is called the *perfect fifth* and is seven semitones higher than the root. If the root has frequency f, then the perfect fifth has frequency $(3/2)f$, a note that sounds beautiful when played together with f (see Unit 1 if you don't remember why). Sometimes, we have additional notes that are octaves of other notes in this chord.

For example, we can make a chord from the root A, the major third C sharp, and the perfect fifth E. Since the root is A, and the pattern consists of a root, a major third, and a perfect fifth, this chord is called A major.

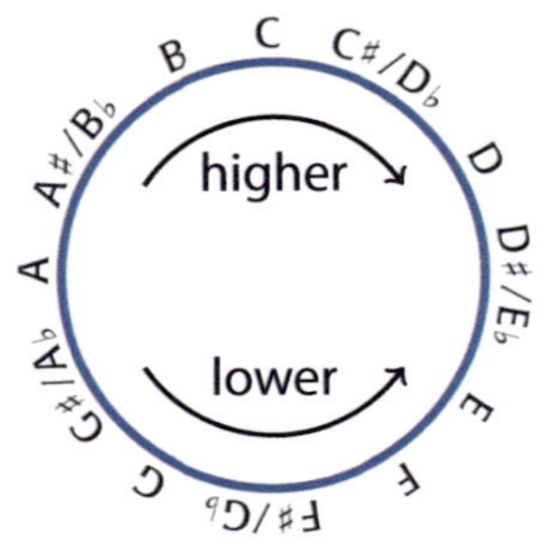

1. What are the three notes that make up a C major chord?

2. What are the three notes that make up a D major chord?

Here's how we can play A major on a ʻukulele. The A note is 2 semitones higher than G so we can put a finger at (10, 9). E is 4 semitones higher than C so we can put another finger at (8, 7). A is 5 semitones higher than E so we can put a finger at (7, 5). And C sharp is 4 semitones higher than A so we can put a fourth finger at (8, 3). Then we can strum the ʻukulele to play the A major chord.

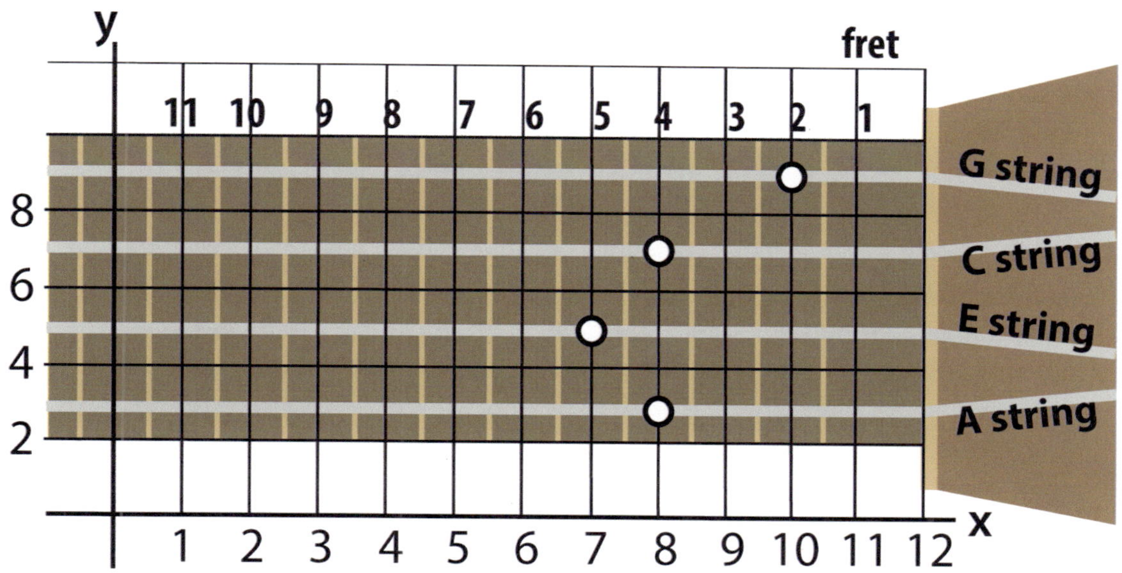

3a. Let's find other ways to play the A major chord. Use the chromatic scale circle to complete the following table.

	Root, A	Major third, C♯	Perfect fifth, E
G string	2nd fret		
C string			4th fret
E string	5th fret		
A string		4th fret	

3b. Using the table from part 3a, find **two** other ways to make the A major chord. Write it as four coordinates—for example (10, 9), (8, 7), (7, 5), and (8, 3). Use 12 as the x-coordinate for open strings—for example, (12, 3) should represent an open A string.
Hint: You have to use all four strings to play all three notes. One string to play more than one note, but two strings can play the same note.

4. A major is made up of the notes A, C sharp, and E. Suppose that we increase each of these notes by 1 semitone to B flat, D, and F. What chord do we have now?

5. What happens when you increase or decrease the notes of a chord by a semitone?

6. We know that one way to play A major is by putting our fingers on (10, 9), (8, 7), (7, 5), and (8, 3) and strumming. Suppose that we moved our hand to $(10+\alpha, 9+\beta)$, $(8+\alpha, 7+\beta)$, $(7+\alpha, 5+\beta)$, and $(8+\alpha, 3+\beta)$ strummed and heard F major. What are the values of α and β?

Activity 4.4 - 'Ukulele Design I

The 'ukulele comes in many different sizes. Here's a picture of a standard 'ukulele, which is also called the soprano 'ukulele.

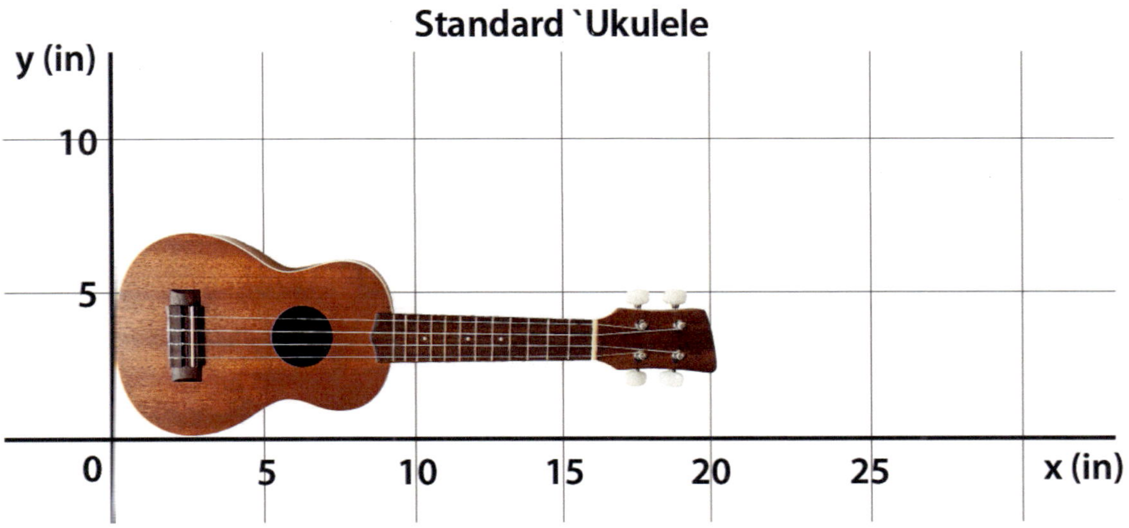

1. We can dilate the standard 'ukulele to the size of a pocket 'ukulele, which is 16 inches long.

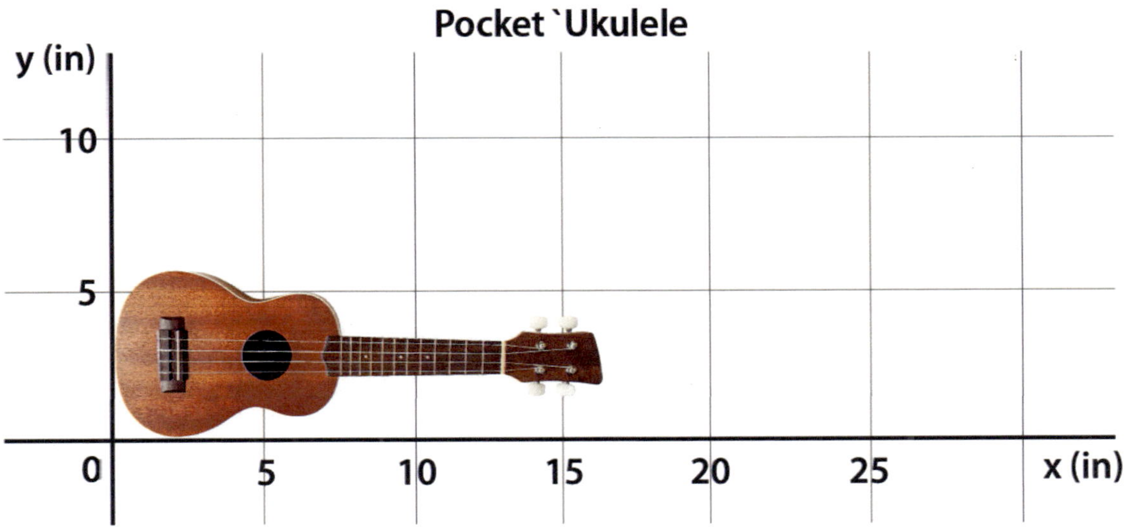

a. What is the scale factor of the pocket 'ukulele compared to the standard 'ukulele?
Hint: scale factor of standard 'ukulele is 1.

b. Why would someone want a pocket 'ukulele instead of a standard 'ukulele? Share your ideas with the class.

2. We can also dilate the standard 'ukulele to the size of a baritone 'ukulele. This baritone 'ukulele is a little bit simplified. We'll talk more about that later.

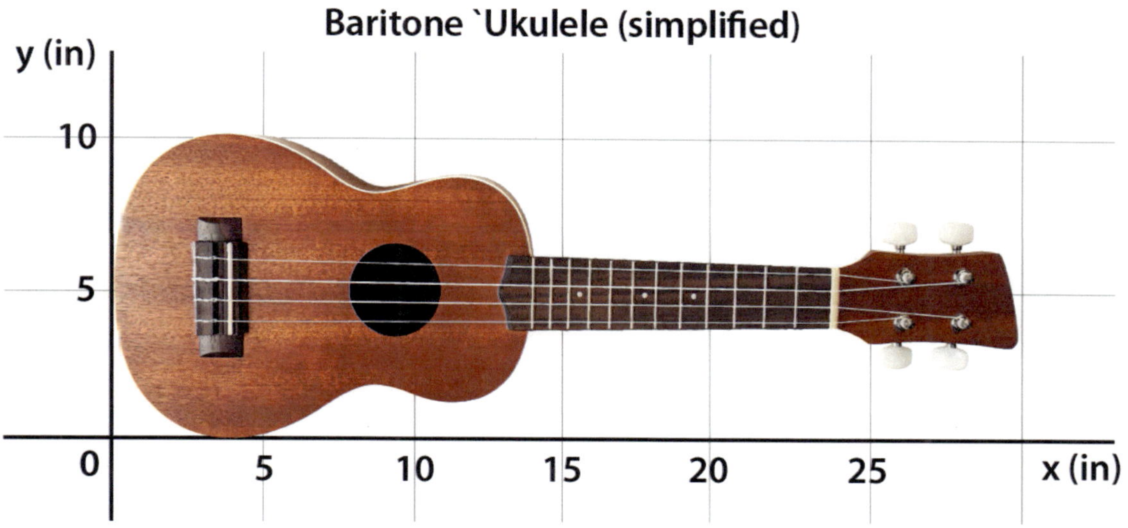

a. What is the scale factor of the baritone 'ukulele compared to the standard 'ukulele? Hint: scale factor of standard 'ukulele is 1.

b. If the strings on the standard 'ukulele are L inches long, how long are the strings on the baritone 'ukulele (in terms of L)?

c. Suppose that the strings on both the standard and the baritone 'ukulele are exactly the same, except for length. In other words, they have the same tightness and thickness, are made of the same material, etc., but they have different lengths. If you plan an open string on the standard 'ukulele and it vibrates at frequency f, what frequency will the same string vibrate at on the baritone 'ukulele?

d. The open strings of a standard 'ukulele play (from top to bottom) G, C, E, and A. What are the notes of the open strings of a baritone 'ukulele?
Hint: If a note has frequency f, then its *perfect fifth* has frequency $(3/2)f$. You might have to review Activity 4.3.

e. This baritone 'ukulele is simplified. A real baritone 'ukulele has more frets. Explain why more frets are needed and support your ideas with the previous answers.

Activity 4.5 - 'Ukulele Design II

The 'ukulele is traditionally a right-handed instrument—you strum and pick with your right hand. Many left-handed players just learn to play the 'ukulele right-handed. Let's look at some of the other options.

1. Some players will just rotate the right-handed 'ukulele 180° and play it upside down. (Notice the order of the strings.)

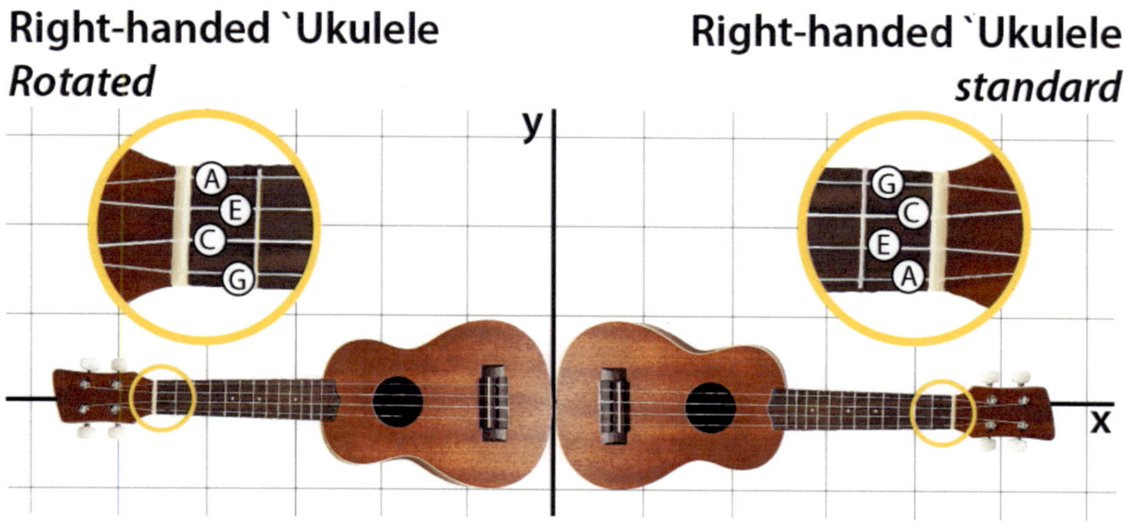

Suppose that a right-handed musician is playing a note by placing their finger at (x, y). If a left-handed musician wants to play the same note on a rotated, right-handed 'ukulele (like the one pictured), where should he/she place their finger?

2. Some players do not want to play the right-handed 'ukulele upside down. So, in addition to rotating the right-handed 'ukulele $180°$, they restring/retune the instrument so the string order is correct.

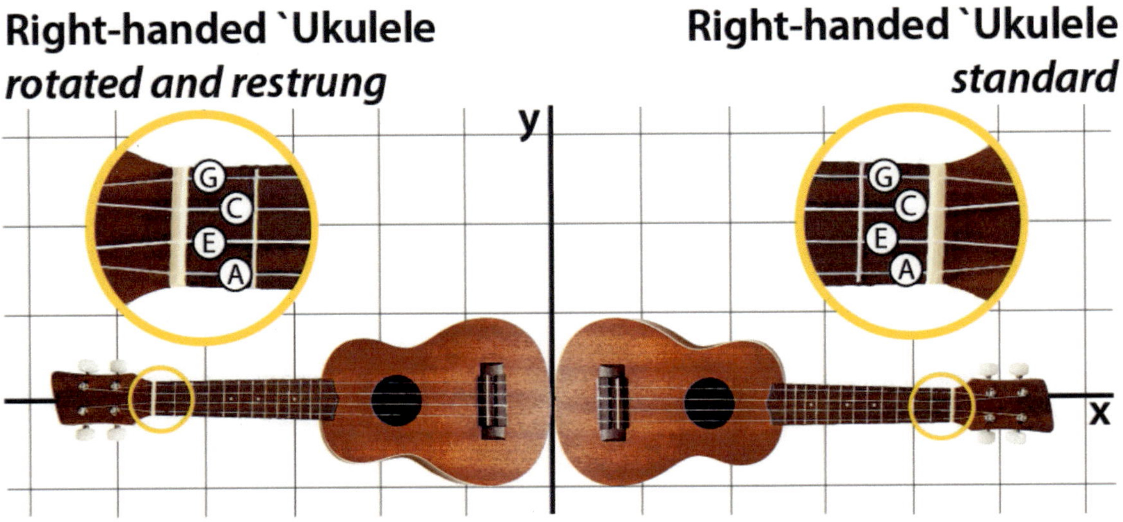

Suppose that a right-handed musician is playing a note by placing their finger at (x, y). If a left-handed musician wants to play the same note on a rotated, restrung, right-handed 'ukulele (like the one pictured), where should he/she place their finger?

3. Some players will spend extra money to buy a left-handed 'ukulele that looks like a mirror reflection of a right-handed 'ukulele.

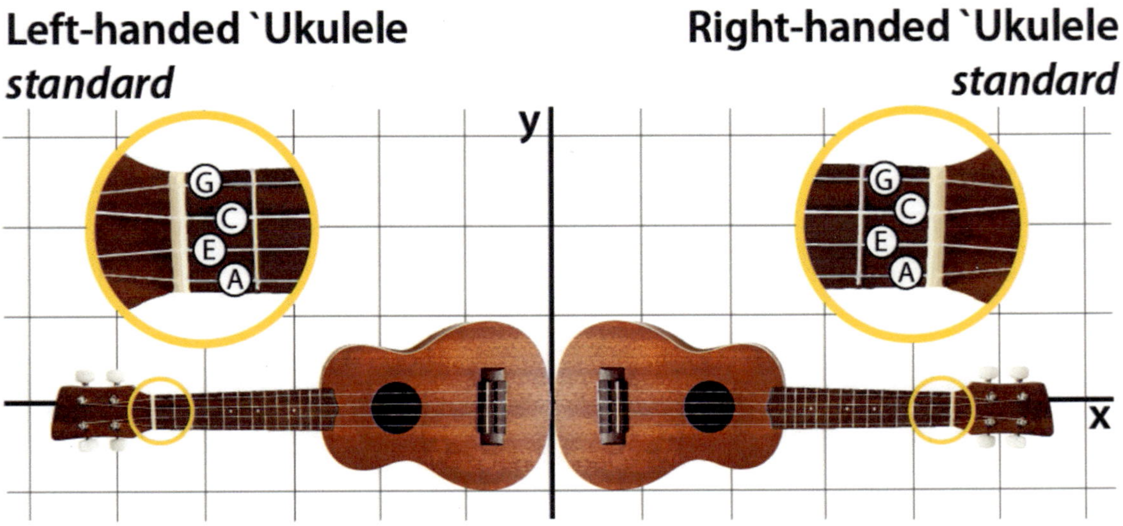

Suppose that a right-handed musician is playing a note by placing their finger at (x, y). If a left-handed musician wants to play the same note on a left-handed 'ukulele (like the one pictured), where should he/she place their finger?

4. So a left-handed player can play a right-handed 'ukulele, a rotated right-handed 'ukulele (like in part 1), a rotated and restrung right-handed 'ukulele (like in part 2), or left-handed 'ukulele (like in part 3). Which would you choose to do? Why? Why is it a better option to you than the others?

Unit 5: Measurement Geometry

Activity 5.1 - Sound reflections I

Sound reflects off of hard surfaces. The angle that sound approaches a surface is the same as the angle that it leaves the surface when it is reflected. For example, if you play a ipu and the sound wave hits a wall at $30°$, then it will reflect away at a $30°$ angle and continue on its way. This is called the Law of Reflection.

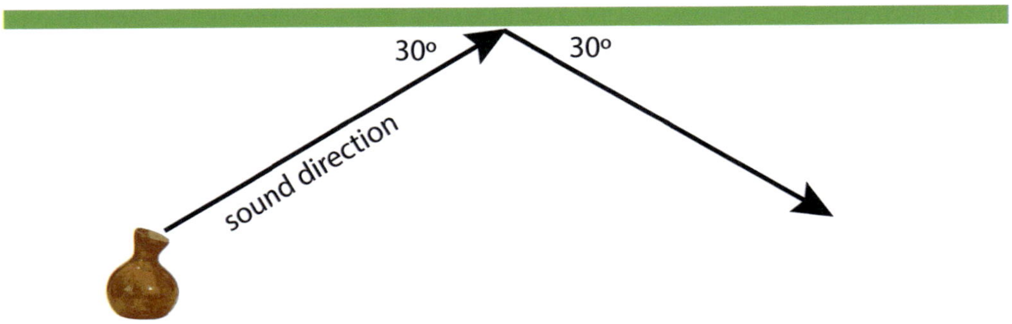

Side note: This is a simplification of how waves move. Remember that waves actually spread out in many directions instead of staying together and moving in one line. Let's just keep it simple for now. Later, we'll look at sound waves moving in different directions.

1. What are the measures of the following angles?

a. Angle a.

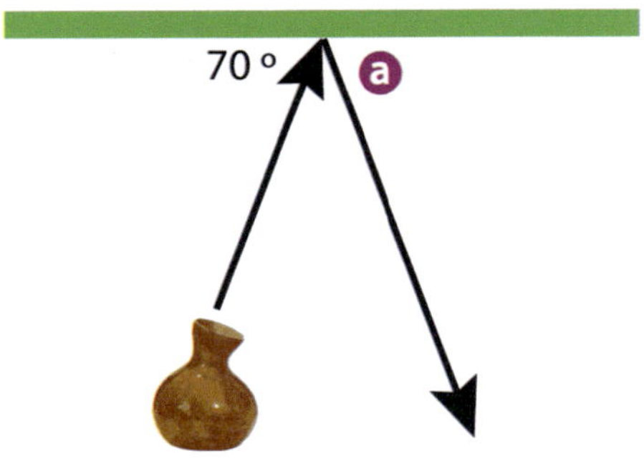

b. Angle b.

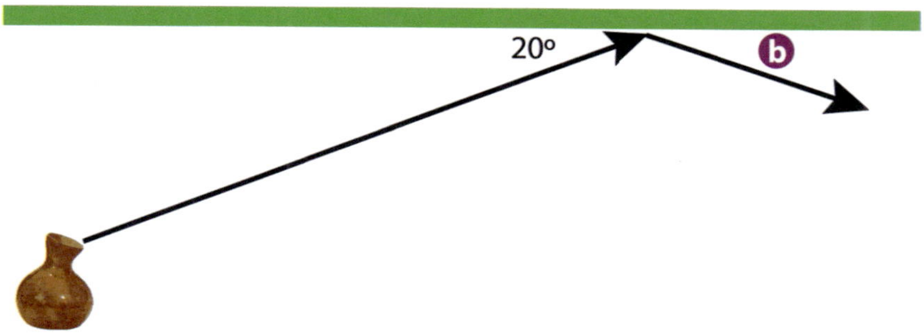

c. Angle c.

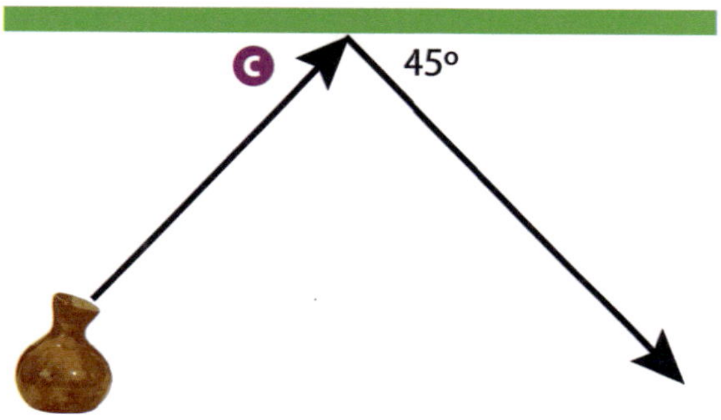

d. Angle d.

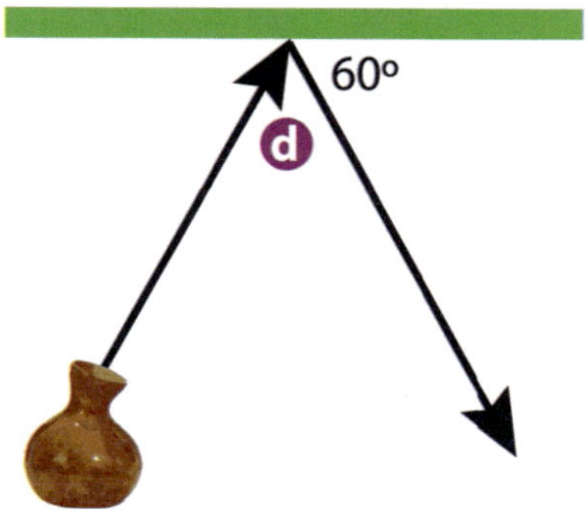

2. This is related to another law called Snell's Law. Look on the internet to see what Snell's Law is about.

a. What are some of the many things that Snell's Law talks about? It can get complicated so be sure to work together and share your ideas with your classmates.

b. Where do you see Snell's Law in real life? Share your ideas in your classroom or on the online comment section .

Activity 5.2 - Sound reflections II

Recall that the Law of Reflection says that when sound approaches a hard surface and is reflected away, the angle that it approaches the surface is equal to the angle that it leaves.

1. When sound wave moves toward a corner with two walls, the sound usually reflects off of both walls before going off in some direction in the room.

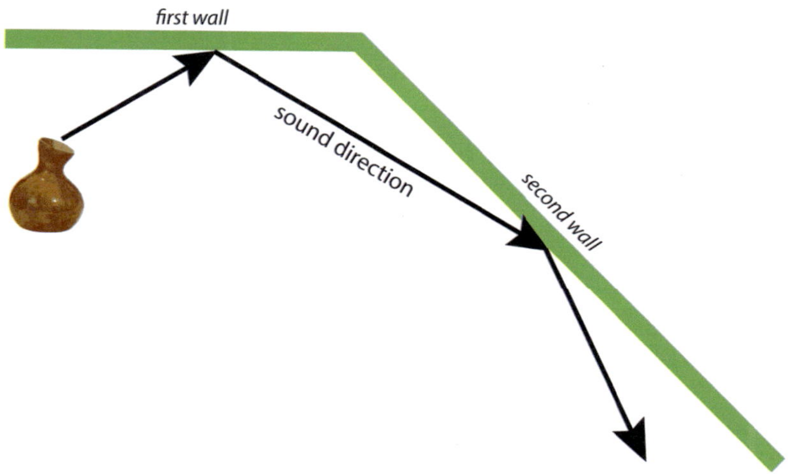

Let's take a look at what happens when the two walls are perpendicular to each other.

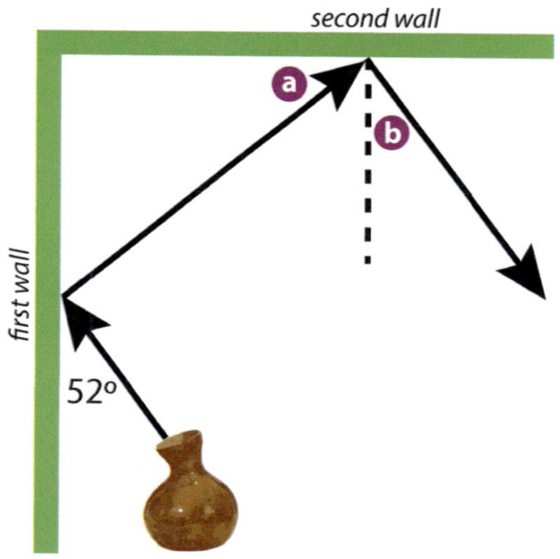

a. Find the measure of angle a. Explain your work, including when you needed to use the Law of Reflection or a math theorem.

b. The dashed line is parallel to first wall. Find the measure of angle b. Explain your work.

c. What can you say about the *direction of the sound wave approaching the first wall* compared to the *direction of the sound wave leaving the second wall*?

d. When designing a *hale* (house or building) for listening to music, architects usually avoid having perpendicular walls in their designs. Why do you think that is?

2. Here's another 90° corner. The dashed line is parallel to the first wall. Angles a and c are congruent. Angles d and e are complementary.

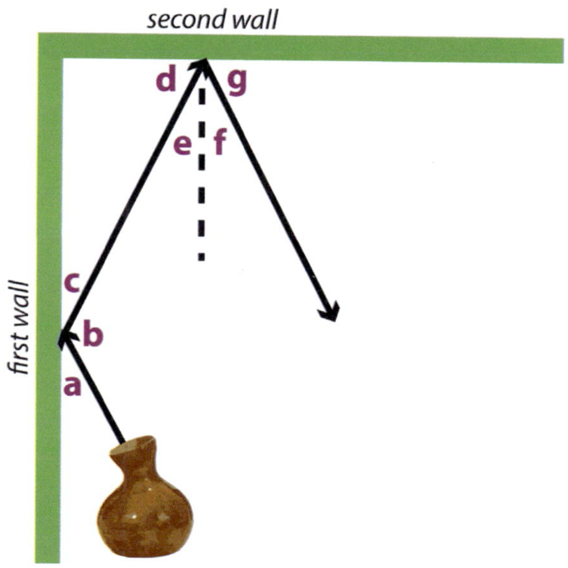

a. What are four other pairs of congruent angles? Write as pairs, for example a & c.

b. What are four other pairs of complementary angles? Write as pairs, for example d & e.

3. Here's a corner that is *not* at a right angle.

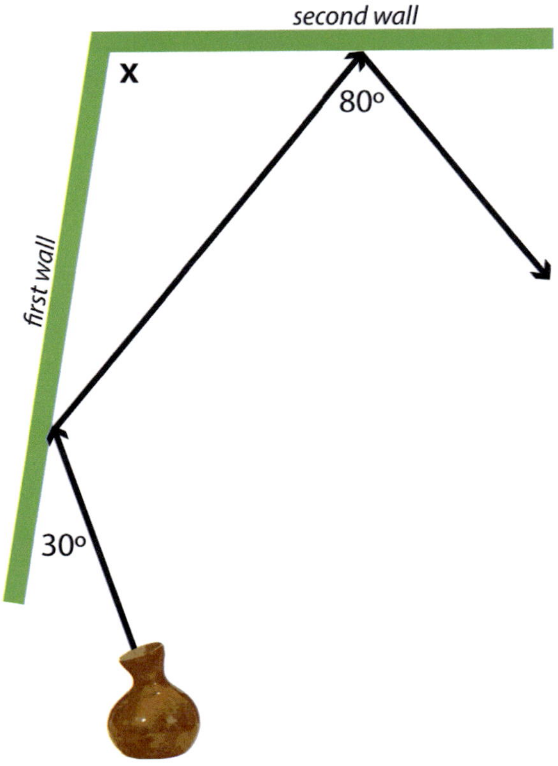

a. What is the measure of the angle of the corner, x?

b. Think about the *direction of the sound wave approaching the first wall* compared to the *direction of the sound wave leaving the second wall.* How does this compare to the perpendicular walls in parts 1 and 2?

Activity 5.3 - Reverberations

For this activity, use 340 m/s for the speed of sound. You may also need to use the Law of Reflection at some point.

Sound waves spread out in many directions. Sometimes we need to understand the distances of things around us and how that distance affects sound.

Imagine that your hula class takes place in a valley. When kumu plays an ipu, some of the sound will go straight to you so you hear it right away. Some of the sound will actually go away from you toward mauka (the mountains). Then the sound waves will reflect off the mountains and come back to you. You might hear the reflected sound right away if the mountains are close or you might hear the sound a couple of seconds later in an echo if the mountains are far away.

1. Suppose that you're at the coordinates (0 m, 0 m), your kumu is at (40.8 m, 54.4 m), and a mountain is at (224.4 m, 299.2 m). You may use a calculator for parts a-g.

a. How far away is your kumu from you?

b. If kumu plays the *pū* (conch shell), how many seconds later will you first hear the sound of the pū? Round to three decimal places if needed.

c. How far away are the mountains from the kumu?

d. How far away are the mountains from you?

e. How far must the sound of the pū travel from the kumu to the mountains and then to you?

f. How how many seconds after the kumu plays the pū, will you hear its sound after it reflects off of the mountains? Round to three decimal places if needed.

g. How much time passes between when you first hear the sound and when you hear its echo from the mountains?

2. If you hear a sound once, then it travels far away, reflects, and you hear it a second time, this is called an *echo*. Since sound spreads in many directions, it is more common to have sounds bouncing all over the place before reaching your ears. So we don't just hear a clear sound twice. Instead we hear a sound many times stretched out over a short duration. This is called *reverberation*. Here's a picture of sound bouncing all over the *ana* (cave) before reaching the listener.

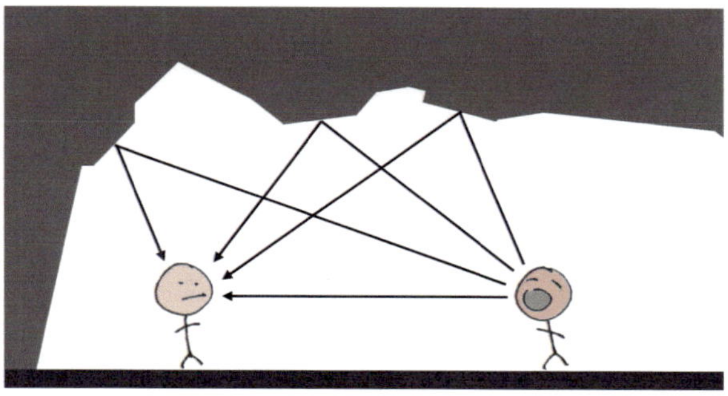

The first sound this listener hears is the sound that went straight to him/her. The last sound is the sound that bounced around the *ana* the furthest before reaching the listener. Roughly speaking, *reverberation time* is the time between when you hear the first sound and when most of the sound goes away.

Suppose that someone is singing 170 m away.

a. How long after the singer sings, does it take for the sound to first reach you? Round to three decimal places.

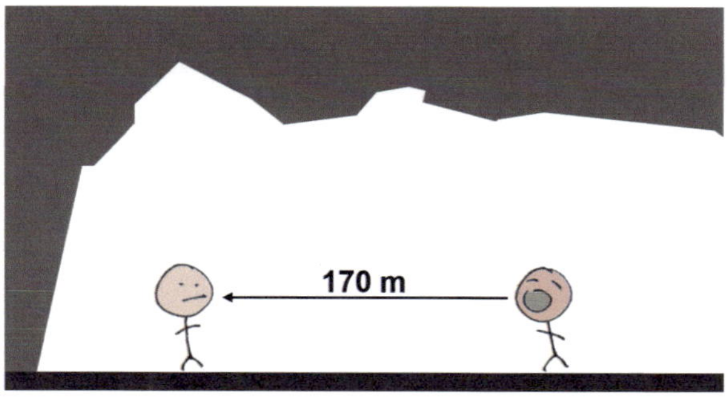

b. Some of the sound does not go straight towards you, but instead bounces around first. If the slowest and last sound traveled 136 m to the ceiling before reflecting another 102 m to you, how long did it take for this sound to reach you? Round to three decimal places.

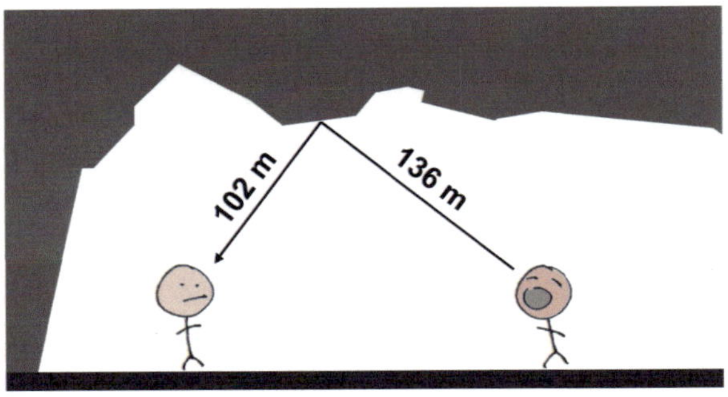

c. Suppose that it's silent after the last sound wave in Part 2b reaches you. What is the reverberation time?

d. To control reverberation, we have to carefully design the angles and shape of the *ana* or where ever you like to listen to music. What is the measure of angle α? Hint: look at how the lines form a triangle. On each side of the triangle, write down how long it took sound to travel that distance.

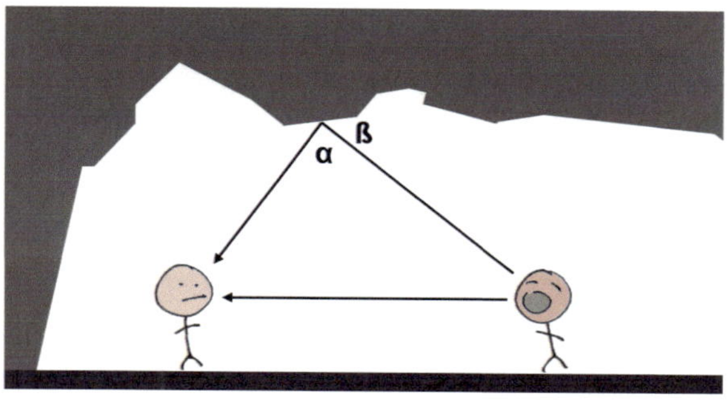

e. Use your answer to Part d and the Law of Reflection to figure out the measure of angle β.

3. Think about times where you've been in a place with a lot or very little reverberation or echo. Describe the environment. What do you think it is about those environments that give it a lot or little reverberation and echo? Share your ideas in your classroom on the online comment section .

Activity 5.4 - Helmholtz Resonator

We know how the strings of a 'ukulele make sound waves. What about an ipu? A ipu works like a Helmholtz resonator, which is any instrument that is made up of a narrow tube connected to a larger hollow body. For example, a milk jug would make a great Helmholtz resonator.

ipu **Helmholtz resonator**

When you hit or blow across a Helmholtz resonator, the air in the neck lightly bounces in and out of the instrument. The air in the neck bounces because the air in the body can be squished and stretched. As the air in the neck bounces, it creates sound waves outside of the instrument.

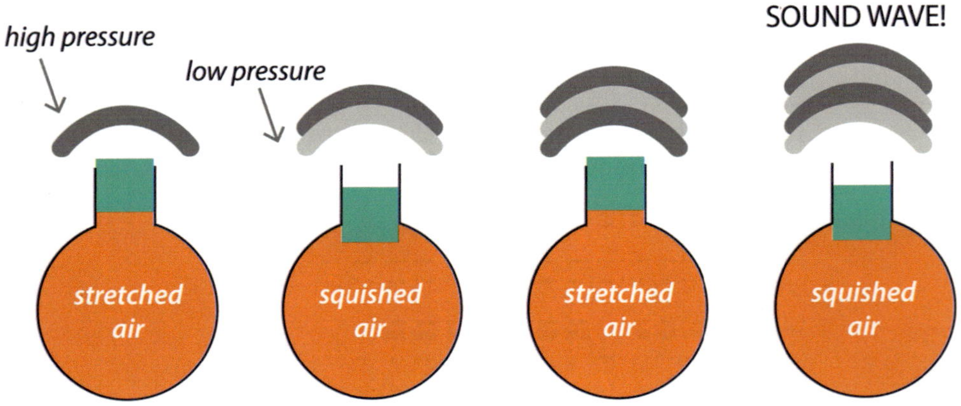

The air in the NECK moves up and down,
while air in the BODY stretches and squishes.
The moving air in the NECK makes a sound wave outside!

1. Suppose that you're playing several ipu in a room. They all have the same neck, but the volume of their body is V. Then the frequency that the ipu make is $f = k\sqrt{1/V}$, where k is a constant positive number.

a. If you increase the volume of the ipu body, will you get a higher or lower sound? How can you tell by looking at the formula for frequency?

b. Rank the following ipu from *lowest* sound to *highest*. Everything expect the volumes of their bodies are the same. Assume that the body is a perfect sphere.
Ipu 1 has body radius 5 in.
Ipu 2 has body volume 900 in^3.
Ipu 3 has body diameter 13 in.
Ipu 4 has circumference of 8π inches at the widest part of the body.

c. What other objects (musical instruments or not), can be a Helmholtz resonator? Explain why and share your ideas with your classmates or on the online comment section.

Unit 6: Statistics

Activity 6.1 - Reverberation Time

Recall that sound waves spread out and bounce around. So sometimes we can still hear a sound a bit after it is no longer being made. If we hear it clearly after a short pause, then it is usually called an *echo*. If we hear it fade away–often without a pause–then it is usually a *reverberation*. As the sound waves spread, it weakens and becomes quieter. The loudness of the sound is called volume and is measured in decibels (dB). This definition of volume should not be confused with the volume that describes the amount of space in an object, which is measured in liters or cubic meters.

Suppose that you're setting up a large room for a Hawaiian music concert. To test if it is set up the way you want, you hit kālaʻau sticks together once (at time 0 ms) and use a machine to record the volume of the sound. You can watch a video about kālaʻau sticks on the book's website (`https://www.stemd2.com`).

1. Here is a graph of the volume in the recording from the first setup.

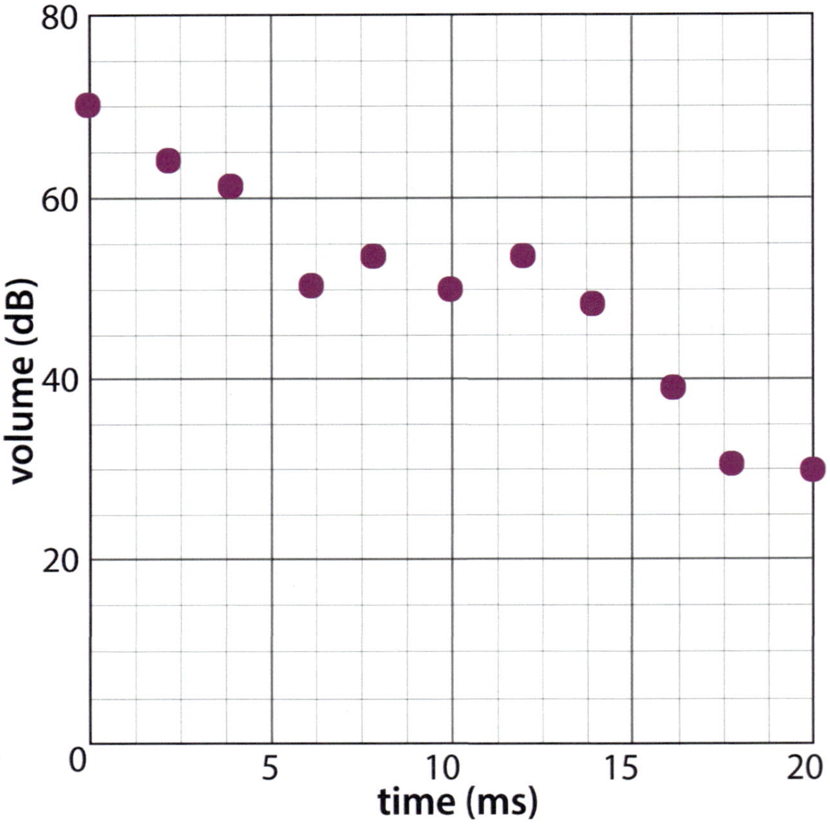

a. Does this scatter plot show a positive association, a negative association, or no association?

b. Describe the real-world reason why your answer to Part 1a makes sense.

c. Assume that there is a linear relationship shown in the graph. What is the equation for this line? Let y be loudness (volume) and t be time.

2. Here is the recording from the second setup.

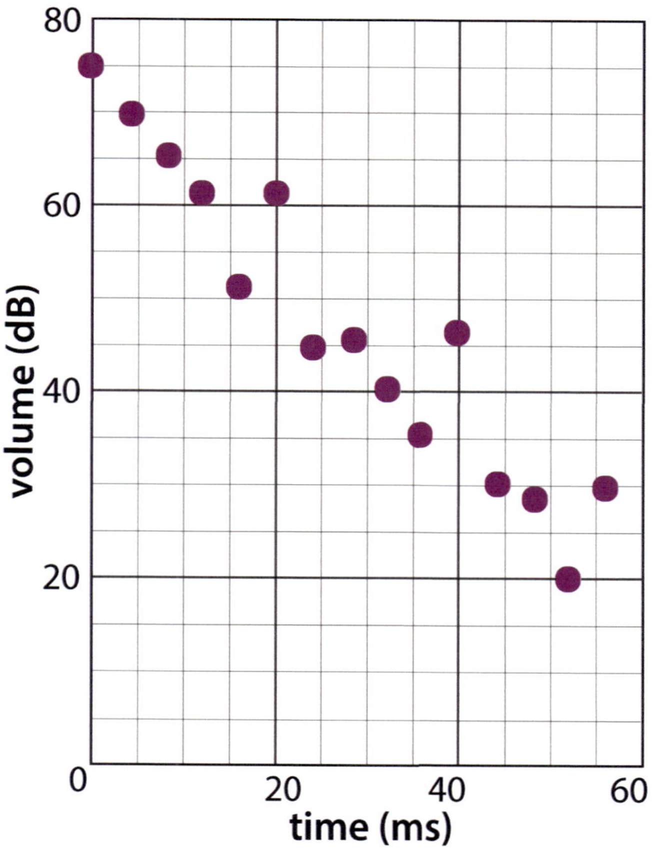

a. Does this scatter plot show a positive association, a negative association, or no association?

b. Describe any clusters that you see on the plot.

c. Describe some of the outliers on the plot.

d. Take a look at the outliers you found in part 2c. Why might there be outliers? Are these readings higher or lower than expected? Why might they be going higher/lower?

e. Assume that there is a linear relationship shown in the graph. What is the equation for this line? Let y be loudness and t be time.

3. *Reverberation time* is the time it takes for a sound's volume to drop 60 dB. For example, if a sound starts at 100 dB, the reverberation time is how long it takes for the sound to dampen to 40 dB.

a. What is the reverberation time of the first setup?

b. What is the reverberation time of the second setup?

c. One of these setups results in more echoes. Which one do you think it is and why did you think it was that one?

d. Which of these setups would you choose for a concert and why? Discuss and share your ideas with other groups and on the on the online comment section ➤.

Activity 6.2 - Music Preferences I

Now that we have the room set up for the concert, let's choose the music. We can do this by asking other students what kind of music they want to hear. Here's the data we've collected by asking 210 students about their music preferences.

Music preference

		Hawaiian music	Jawaiian music	TOTAL
Age	10-12	52	78	130
Age	13-15	52	28	80
	TOTAL	104	106	210

1. Let's take a look at each age group separately.
Next to letters A-G, write the relative frequency compared to the total of that **row**. Here, two boxes in the first row are filled as examples. Give your answers as a percentage and round to the nearest whole number. You may use a calculator for this part.

Music preference

		Hawaiian music		Jawaiian music		TOTAL	
Age	10-12	52	40%	78	A	130	100%
Age	13-15	52	B	28	C	80	D
	TOTAL	104	E	106	F	210	G

2. Let's take a look at each music genre separately.
Next to letters A-G, write the relative frequency compared to the total of that **column**. Here, two boxes in the Hawaiian music column are filled as an example. Give your answers as a percentage and round to the nearest whole number. You may use a calculator for this part.

Music preference

Age		Hawaiian music		Jawaiian music		TOTAL	
Age	10-12	52	50	78	A	130	B
Age	13-15	52	C	28	D	80	E
	TOTAL	104	100%	106	F	210	G

3.a. Explain the meaning of the percentages you wrote in Part 1. What do they tell you?

b. Explain the meaning of the percentages you wrote in Part 2. What do they tell you?

4. Of the types of songs that you're thinking of playing for the concert, what percent of them should be Hawaiian songs? What percent of them should be Hip Hop? How did you use the data from the table to help you make your decisions?

Activity 6.3 - Music Preferences II

Your kumu noticed that not everyone voted. Maybe some students wanted to hear other kinds of music. To prepare for the next concert, we let students vote for "other music." Here are the results.

Music Preference

Age		Hawaiian	Jawaiian	Other	TOTAL
	Age 10-12	55	104	21	180
	Age 13-15	57	15	8	80
	TOTAL	112	119	29	260

1. Let's take a look at all the votes.

Next to the letters A-I, write the relative frequency compared to the total number of votes (260). Give your answers as a percentage and round to the nearest whole number. Three boxes are already filled. For example, we can see that 21% of all votes were from 10-12 year old students who wanted to hear Hawaiian music. You may use a calculator for this part.

Music Preference

Age		Hawaiian		Jawaiian		Other		TOTAL	
	Age 10-12	55	21%	104	A	21	B	180	C
	Age 13-15	57	D	15	E	8	F	80	31%
	TOTAL	112	G	119	H	29	I	260	100%

2. What are some conclusions or statements that you can make about the data? For example, can you say that certain groups have a stronger preference for certain kinds of music? If you cannot make any conclusions, why not?

3. Share your ideas with your classmates and online. Did you see any conclusions from other groups that you disagree with? How is it possible for two people to look at the same data and come up with very different conclusions? This actually happens a lot. Talk with your other classmates about how or why this is happens.

Made in the USA
Monee, IL
05 November 2026

67381722R00055